"Are you tough enough to hike this trail?" Dad asked.

"I am!" answered Rozell.

"I'm tough enough too!" added Suzann.

"It's a tough trail," Mom said. "It's not the little trail we took last July. I know this backpack trip will be tough. It could get a little rough!"

"I'm strong," said Rozell.
"It won't be too rough for me!
I'm a tough kid. And I know
how to backpack. I have done
that before!"

Suzann said, "We carry our backpacks. Then we set up our camp. Before long, it's time to eat. We cook and eat outside."

"No hotel rooms for us!"
said Rozell. "No motel beds!
Sleeping bags are good enough
for us. Our room is a tent."

"It's a tough trek," said Dad.
"The hike is rough and quite
long. Before we go, we should
study this map."

"We are tough enough," said Suzann. "This will be our best July trip! It will be superb! Don't you think so, Snap?"

"I wish it were July now,"
said Rozell. "And so does Snap!
He's been on a hike before. He's
tough enough to come with us!"

The End